SPACE WONDERS

INNER PLANETS

BLACK RABBIT BOOKS

TABLE OF CONTENTS

1 Close to the Sun

There are many objects in the night sky. Some of the brightest are planets. Our **solar system** has eight planets. Four of them are closer to the sun than the others. These are called the inner planets.

The inner planets have some things in common. All have a rocky surface. They have a metal center. All have short **orbits**. None of them have rings. In other ways, the inner planets are very different from each other.

Think About It

Why do you think the inner planets are alike?

2

Mercury

Mercury is the smallest planet in our solar system. It is a little bigger than Earth's Moon. Mercury is also the closest to the sun. It zips around the sun in just 88 days.

Mercury has crazy temperatures. During the day, it can be as hot as 800 degrees Fahrenheit (430 degrees Celsius). But Mercury has no **atmosphere**. The planet can't hold in heat. At night, it drops to -290°F (-180°C).

Mercury's surface is covered with **craters**. It has been hit many times by space rocks.

Did You Know?
Mercury is named after the fastest Roman god.

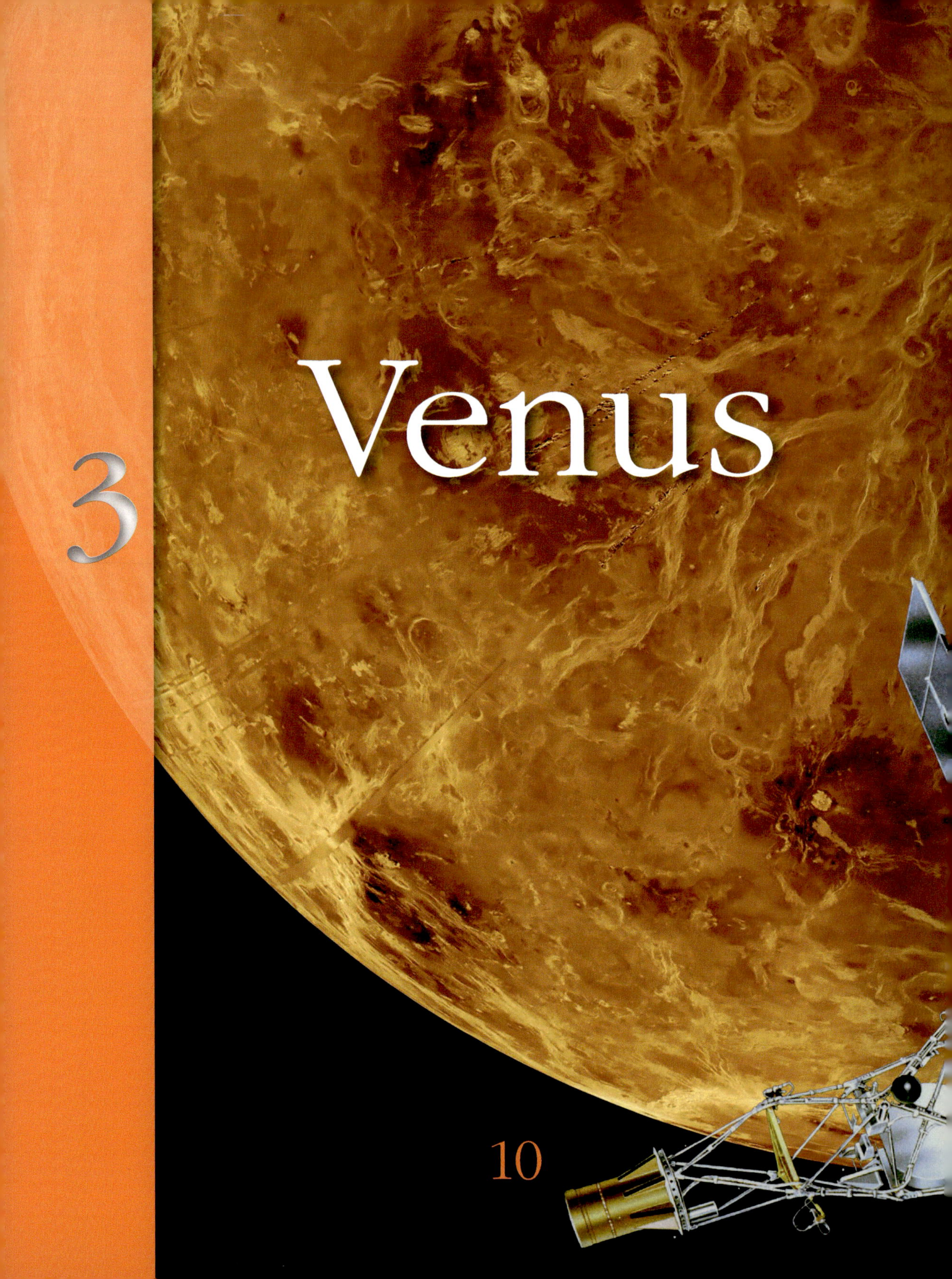

3

Venus

Venus is not the closest planet to the sun. But it is the hottest. Venus is covered with a thick atmosphere. It holds in heat. The temperature is always around 870°F (465°C).

Mariner 2 was the first spacecraft to visit another planet. It flew past Venus in 1962. In 1970, the Russian craft *Venera* flew to Venus. It was the first to land there. Since then, other spacecraft have landed or flown past.

Mariner 2

Did You Know?

Venus spins slower than it orbits. A year on Venus is shorter than a day.

4 Earth

Earth is our home planet. It is third from the sun. It is not too hot or too cold. Scientists call this the Goldilocks Zone. It is the perfect place for life to grow.

Earth's atmosphere protects us from the sun's **radiation**. It traps heat to warm the planet. It also protects us from space rocks.

Earth has four **layers**. The middle layers are made of metal. The mantle is hot, liquid rock. The crust is the surface of the planet.

Did You Know?
About 71 percent of the Earth's surface is covered by water.

Mars

5

Perseverance

Mars is named after the Roman god of war. Mars is also called the Red Planet. That's because the surface looks red. It is full of iron. Mars is the closest planet to Earth.

Many spacecraft have explored Mars. In 2021, *Perseverance* landed there. This rover carried a helicopter called *Ingenuity*. It made the first powered flights on Mars. *Ingenuity* flew for three years. It scouted places to explore. This helped NASA plan future missions.

Think About It

Do you think people will ever live on Mars?

Small Worlds

6

The inner planets formed billions of years ago. Scientists think a giant star exploded. Dust and gases spread. Gravity pulled these together. They formed stars and planets. The inner planets may be made of the same materials as the sun and stars.

We are lucky that we can study the inner planets. When we do, we learn about our neighbors in space. We also learn more about Earth.

Think About It

What are some ways scientists study the inner planets?

MORE TO EXPLORE

FANTASTIC FACTS

Mariner 10 flew past Mercury three times between 1974 and 1975.

Gravity is weaker on Mars than on Earth. You could jump three times higher on Mars.

One canyon on Mars is 10 times as big as the Grand Canyon. If it was on Earth, it would stretch from New York to California!

In 2022, the *Curiosity* rover found signs there might have been an ocean on Mars long ago.

The *Perseverance* rover is the size of a car.

Venus spins backward. The sun rises in the west and sets in the east there.

MORE TO EXPLORE

COOL COMPARISONS

How big are the inner plants?
Compare their sizes across.

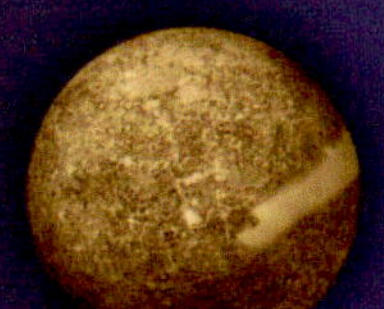

Mercury
1,516 miles (2,440 km)

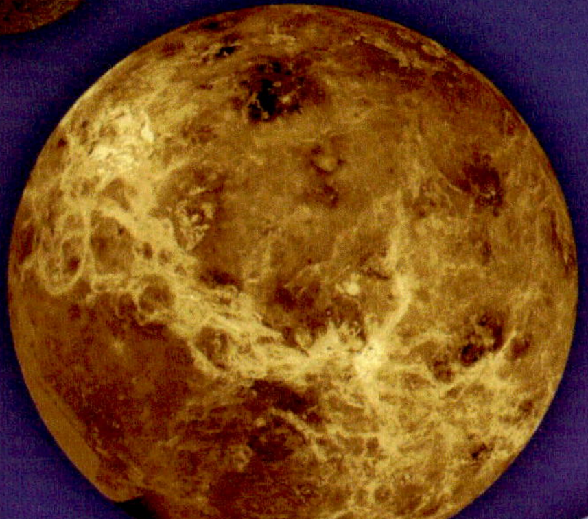

Venus
3,760 miles (6,052 km)

Earth
3,959 miles (6,371 km)

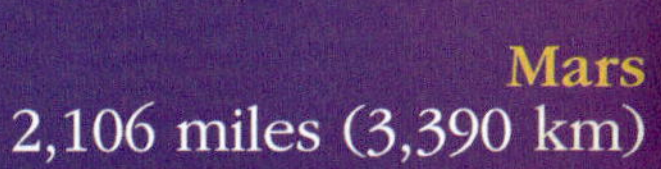

Mars
2,106 miles (3,390 km)

MORE TO EXPLORE

RESOURCES

Glossary

atmosphere (AT-muhs-feer) The gases that surround a planet.

crater (KREY-ter) A large round hole in the ground made by something falling from the sky.

layer (LEY-er) An amount of something that is spread over an area.

orbit (AWR-bit) The path taken by one body circling around another body.

radiation (rey-dee-EY-shuhn) Energy that comes from a source in the form of waves or rays you cannot see.

solar system (SOH-luhr SYS-tum) The planets and space objects that circle the sun.

Read More

Betts, Bruce. *Mars: The Red Planet*. Minneapolis: Lerner Publications, 2025.

Leaf, Christina. *The Inner Planets*. Minneapolis: Bellwether Media, 2023.

Index

TOP RANK is published by Black Rabbit Books, P.O. Box 227, Mankato, MN, 56002.

• Top Rank is an imprint of Black Rabbit Books • Series designed by Danny Nanos • Book designed by Jason Knudson • Photographs © Freepik/alexkoral, 2–3, 19; NASA, 15; NASA/JPL-Caltech, 16–17, 17; Science Source/Ron Miller, 12; Shutterstock/B-E, 9, buradaki, cover, 6–7, 20, ConceptCafe, 18, DestinaDesign, 14, Jurik Peter, 8–9; Wikimedia Commons/Jack, 2, 5, 23, NASA, 19, NASA/JPL, 4, 10–11, Rimaakter45, 13 • Printed in the United States of America

Library of Congress Cataloging-in-Publication Data: Names: Mattern, Joanne, 1963- author. | Title: Inner planets / Joanne Mattern. | Description: Mankato, MN: Black Rabbit Books, [2026] | Series: Space wonders | Includes bibliographical references and index. | Audience: Ages 8–11 | Audience: Grades 2–3 | Identifiers: LCCN 2025011954 (print) | LCCN 2025011955 (ebook) | ISBN 9781645825104 (library binding) | ISBN 9781645825289 (paperback) | ISBN 9781645825463 (ebook) | Subjects: LCSH: Inner planets—Juvenile literature. | Classification: LCC QB606 .M38 2026 (print) | LCC QB606 (ebook) | DDC 523.4—dc23/eng/20250324 | LC record available at https://lccn.loc.gov/2025011954